Math

Grade 8

This book has been designed to help all students succeed in math. It is part of a *Basics First* series that provides students with opportunities to practice basic math skills that will help them begin to understand many of the mathematical properties that they will use over and over throughout their lives. The activities can be used as supplemental material to reinforce any existing math program.

The activities in this book have been created to help students feel confident in math computation. In order to ensure success, students must be guided through the activities presented. Most of the activities feature an amusing illustration which will motivate students and help maintain a high level of interest as students complete them. Allow students to use models at any time while working on the activities in this book.

A variety of fun and simple formats are included throughout the book. Students will enjoy working with menus, "shopping" at sales, doing crossword puzzles, balancing checking accounts, and much more.

The skills covered in this book can be taught in the classroom or at home. They include positive and negative rational numbers; solving multiple-step equations; using fractional percents; the Pythagorean theorem with estimating square roots; solving right triangles for area and perimeter; area and perimeter for trapezoids; finding area and circumference for circles; finding volume and surface area for square pyramids, tetrahedrons, cylinders, cones, and spheres; constructions; and charting and graphing first-degree functions.

This book's level of difficulty is geared toward the average math student. However, in some areas, challenge questions have been included and are indicated by the symbol, (*).

Integer Innings

Multi-step integer problems

Name ______________________________

For a home run, start at home plate and solve each problem to 1st, 2nd, 3rd, and then back to home.

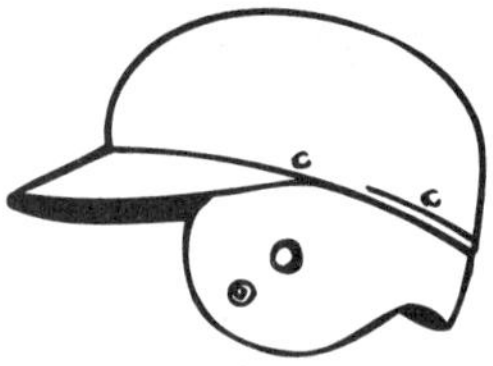

$$\text{2nd} = \frac{^{-}12 + (^{-}9 - {^{-}6})}{^{-}5}$$

$$\text{3rd} = \frac{^{-}6 - [^{-}3(^{-}15 + 5)]}{(^{-}2 + {^{-}7})}$$

$$\text{1st} = \frac{^{-}8(^{-}3 + {^{-}5})}{(^{-}1 + {^{-}3})}$$

$$\text{Home} = \frac{[7(^{-}10 - {^{-}2}) - 4]}{^{-}2(^{-}5)}$$

A. 1st = ________

2nd = ________

3rd = ________

Home = ______

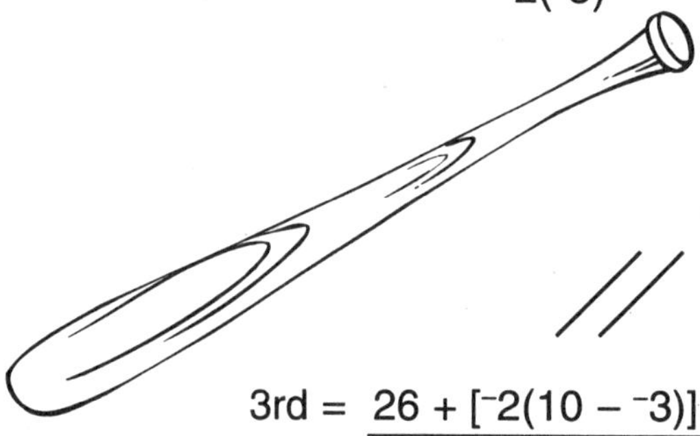

B. 1st = ________

2nd = ________

3rd = ________

Home = ______

$$\text{2nd} = \frac{(^{-}12 + 19) - {^{-}8}}{(^{-}1 - 4)}$$

$$\text{3rd} = \frac{26 + [^{-}2(10 - {^{-}3})]}{(^{-}3 \bullet 12)}$$

$$\text{1st} = \frac{9(^{-}6 - 2)}{^{-}1(^{-}36)}$$

$$\text{Home} = \frac{(^{-}8 + 3)(^{-}2 + {^{-}3})}{(10 - 15)(^{-}9 + 8)}$$

$$\text{2nd} = \frac{[19 + (^{-}3 \bullet 4)] - {^{-}8}}{5(^{-}3)}$$

$$\text{3rd} = \frac{\left(\frac{100}{^{-}10}\right) + 15}{(5 - 7)}$$

$$\text{1st} = \frac{149 - (^{-}4 \bullet {^{-}7})}{^{-}11(11)}$$

$$\text{Home} = \frac{(^{-}3 + {^{-}3})(^{-}2 \bullet 3)}{^{-}6}$$

C. 1st = ________

2nd = ________

3rd = ________

Home = ______

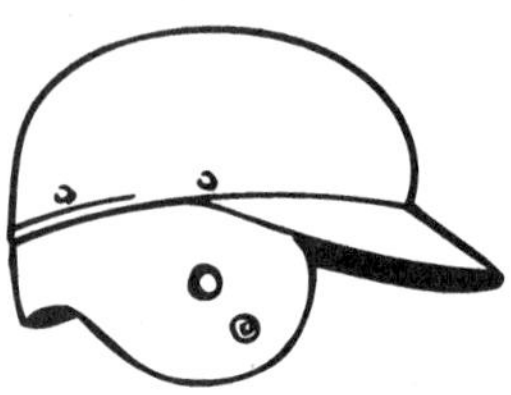

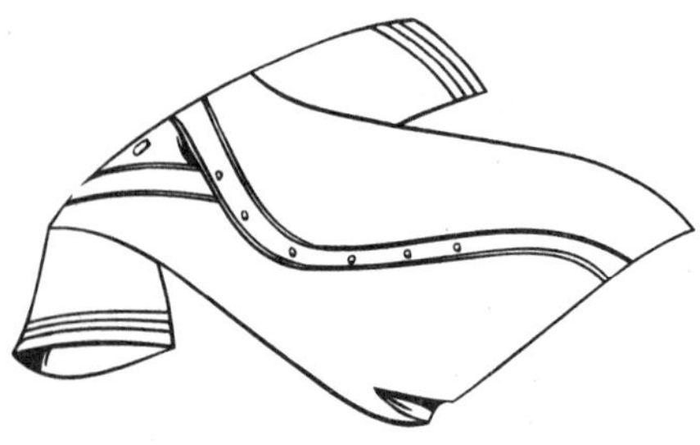

"May I Take Your Order?"

Ordering fractional integers

Name ______________________________

Put each set of numbers below in order from greatest to smallest.

I. $-1\frac{1}{3}$, 0, $2\frac{2}{3}$, $1\frac{3}{4}$, $-\frac{5}{8}$, $-\frac{2}{3}$, $1\frac{3}{5}$, $2\frac{5}{7}$, $-1\frac{3}{8}$, $-\frac{4}{7}$

1 =

2 =

3 =

4 =

5 =

6 =

7 =

8 =

9 =

10 =

II. $-\frac{5}{16}$, $1\frac{2}{3}$, $\frac{5}{9}$, 0, $-\frac{4}{9}$, $1\frac{3}{5}$, $-1\frac{1}{3}$, $-\frac{3}{8}$, $-1\frac{2}{5}$, $\frac{1}{2}$

1 =

2 =

3 =

4 =

5 =

6 =

7 =

8 =

9 =

10 =

III. $-1\frac{1}{2}$, $\frac{4}{5}$, $-2\frac{3}{16}$, $-1\frac{9}{20}$, $\frac{8}{9}$, $-2\frac{1}{4}$, 0, $\frac{13}{16}$, $-2\frac{2}{9}$, $\frac{7}{8}$

1 =

2 =

3 =

4 =

5 =

6 =

7 =

8 =

9 =

10 =

Hip-hop Fractions

Adding positive and negative fractions

Name ______________________________

Start with problem 1 and calculate each pair of numbers. The answer to the first pair becomes the first number in the second pair, and so on, until the cycle is complete.

A.

2. $\_\_\_\_ + {}^{-}8\frac{7}{8} = \_\_\_\_$

1. ${}^{-}3\frac{3}{4} + {}^{-}5\frac{3}{5} = \_\_\_\_$

3. $\_\_\_\_ + 2\frac{1}{2} = \_\_\_\_$

6. ${}^{-}3\frac{3}{4} = \_\_\_\_ + \_\_\_\_$

4. $\_\_\_\_ = 6\frac{11}{12} + \_\_\_\_$

5. $\_\_\_\_ = \frac{-7}{8} + \_\_\_\_$

B.

2. $\_\_\_\_ + {}^{-}8\frac{3}{4} = \_\_\_\_$

1. ${}^{-}2\frac{2}{3} + 2\frac{1}{2} = \_\_\_\_$

3. $\_\_\_\_ + {}^{-}3\frac{2}{3} = \_\_\_\_$

6. ${}^{-}2\frac{2}{3} = \_\_\_\_ + \_\_\_\_$

4. $\_\_\_\_ = 5\frac{5}{8} + \_\_\_\_$

5. $\_\_\_\_ = {}^{-}1\frac{1}{2} + \_\_\_\_$

Fractured Fractions

Subtracting positive and negative fractions

Name ____

Solve each problem below.

1. $^{-}8\frac{1}{3} - {}^{-}5\frac{5}{6} =$ ______

2. ______ $= 2\frac{2}{3} - 5\frac{11}{12}$

3. $^{-}9\frac{3}{4} - 6\frac{5}{12} =$ ______

4. ______ $= {}^{-}8\frac{1}{4} - 7\frac{5}{7}$

5. $^{-}5\frac{7}{18} - {}^{-}12\frac{3}{4} =$ ______

6. $30\frac{3}{4} - {}^{-}5\frac{5}{6} =$ ______

7. $^{-}16 - {}^{-}5\frac{5}{8} =$ ______

8. $5\frac{4}{7} - 12 =$ ______

9. $^{-}12\frac{2}{3} - 15\frac{5}{18} =$ ______

10. $8\frac{7}{8} - 15\frac{5}{6} =$ ______

11. $^{-}10\frac{2}{15} - {}^{-}8\frac{11}{12} =$ ______

12. $^{-}9\frac{1}{4} - 8\frac{4}{5} =$

13. $16\frac{11}{12} - 20\frac{3}{8} =$ ______

14. $^{-}17 - 12\frac{2}{3} =$ ______

Hex-a-Fraction

Adding and subtracting positive and negative fractions

Name ______________________________

Complete the hexagon fraction diagrams below by finding the missing sums or addends. Simplify all answers when possible.

1.

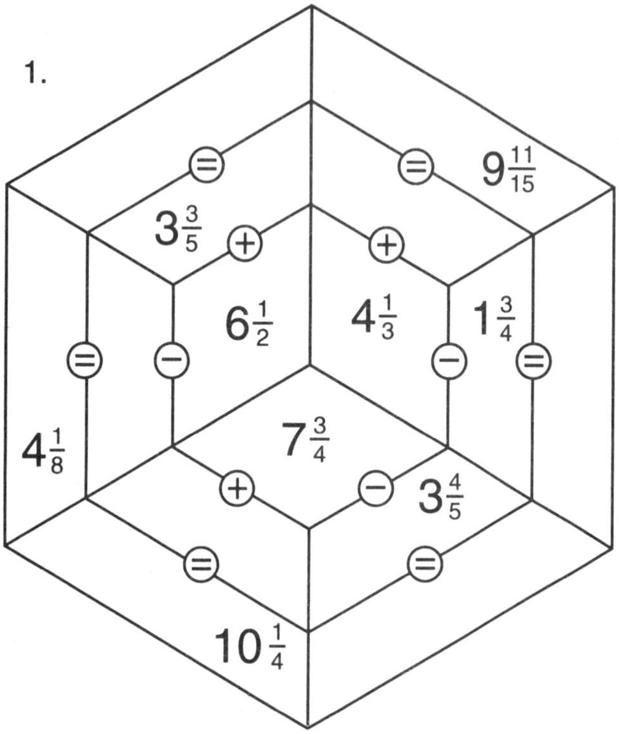

2.

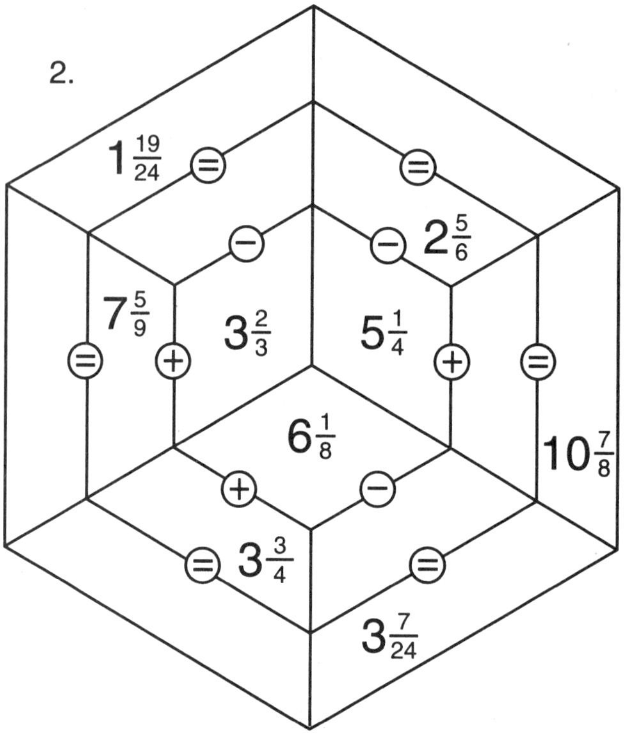

3.

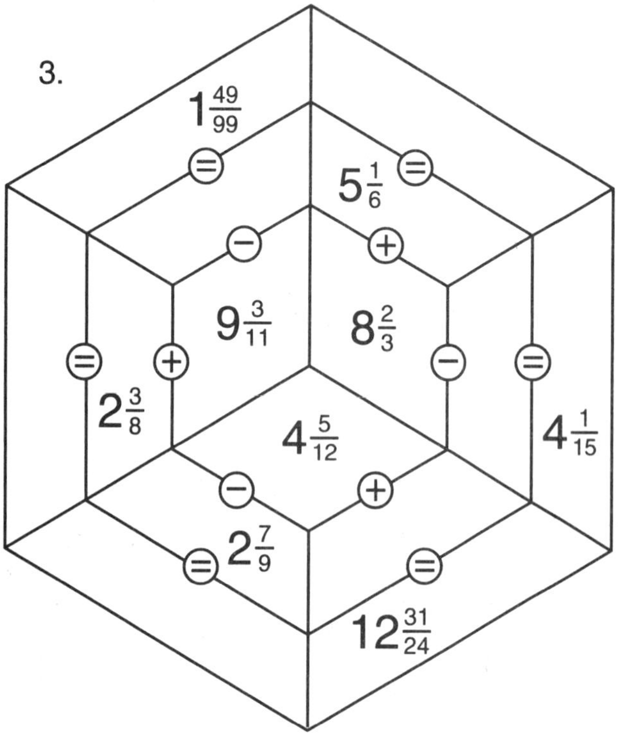

4.

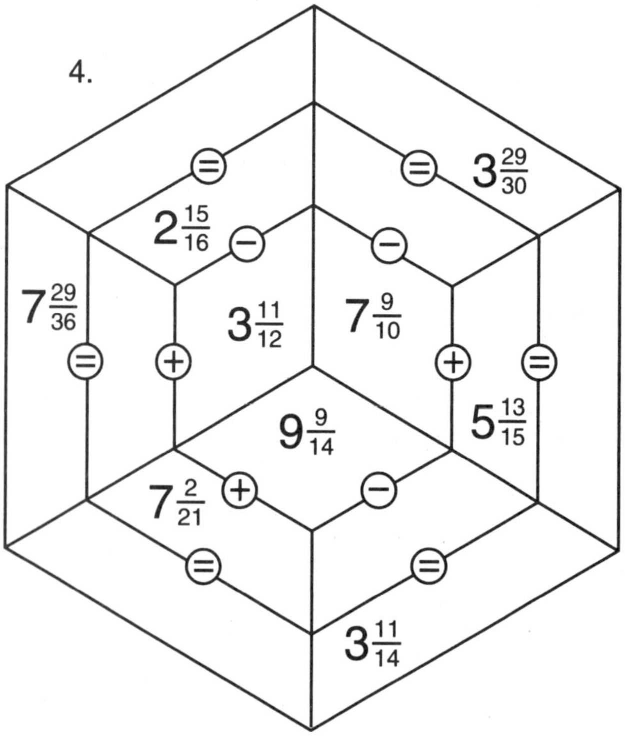

X's and O's

Multiplying positive and negative fractions

Name ____________________

Solve each problem below.

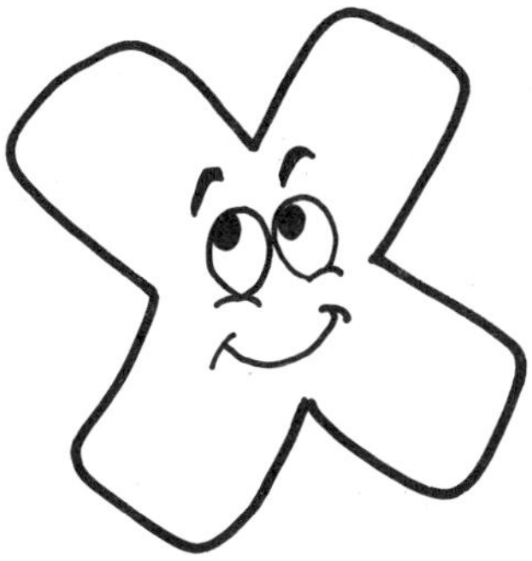

1. $-\frac{3}{4} \times \frac{16}{21} =$

2. $-\frac{8}{9} \times \frac{3}{8} =$

3. $(-\frac{3}{4})^2$

4. $-\frac{5}{6} \times -\frac{4}{15} \times -\frac{24}{25} \times \frac{20}{36} =$

5. $(-\frac{2}{5})^3 =$

6. $(-\frac{1}{4})^4 =$

7. $1\frac{1}{2}(-3\frac{1}{3}) =$

8. $-3\frac{3}{4}(-5\frac{1}{3}) =$

9. $-9\frac{4}{9} \times 1\frac{1}{2} \times -3\frac{3}{5} =$

10. $-6\frac{2}{3} \times 2\frac{2}{5} =$

11. $-5\frac{1}{5} \times 1\frac{7}{13} \times 2\frac{2}{3} =$

12. $(-1\frac{2}{3})^2 =$

13. $-3\frac{2}{3} \times 2\frac{1}{22} \times -2\frac{2}{9} =$

14. $-4\frac{4}{5} \times -1\frac{1}{6} \times -1\frac{1}{14} =$

15. $(-2\frac{3}{4})^3 =$

16. $(-2\frac{1}{2})^4 =$

Monkey Puzzle Tree

Dividing positive and negative fractions

Name ______________________________

Solve each problem below. Then use the code to see what the monkey found at the top of the tree.

CODE
s = $\pm 2\frac{1}{4}$
c = $\pm 1\frac{7}{8}$
n = $\pm 1\frac{1}{2}$
f = ± 6
b = $\pm 1\frac{1}{4}$
o = $\pm 5\frac{5}{9}$
h = $\pm 1\frac{2}{7}$
a = $\pm\frac{1}{2}$
u = ± 3

1. $\dfrac{\frac{4}{5}}{1\frac{3}{5}} =$

2. $3\frac{3}{4} \div {}^-3 =$

3. ${}^-10\frac{2}{3} \div 3\frac{5}{9} =$

4. ${}^-10\frac{1}{8} \div {}^-6\frac{3}{4} =$

5. $\dfrac{5}{{}^-2\frac{2}{3}} =$

6. ${}^-9\frac{2}{7} \div {}^-7\frac{2}{9} =$

7. ${}^-8\frac{1}{3} \div 1\frac{1}{2} =$

8. $6\frac{2}{3} \div {}^-1\frac{1}{9} =$

9. $\dfrac{-\frac{5}{8}}{-\frac{1}{2}} =$

10. $\dfrac{\frac{1}{4}}{\frac{1}{2}} =$

11. $\dfrac{4}{{}^-2\frac{2}{3}} =$

12. ${}^-3\frac{3}{4} \div {}^-7\frac{1}{2} =$

13. $\dfrac{-\frac{3}{4}}{\frac{1}{2}} =$

14. ${}^-5\frac{1}{2} \div 11 =$

15. $1\frac{1}{2} \div {}^-\frac{2}{3} =$

__   __ __ __ __ __   __ __   __ __ __ __ __ __ __

1   2 3 4 5 6   7 8   9 10 11 12 13 14 15

Here's a Letter, Mr. Postman

Adding rational integers

Name ____________________________________

Consider the positive numbers below "checks" deposited to your checking account balance. Consider the negative numbers "bills," decreasing your balance. Find your balance at the end of each work week.

Week 1

Monday: beginning balance: $ 103.67
Today, the postman brings a check for 43.16, and bills for $^{-}197.50$ and $^{-}25$.

Tuesday: beginning balance: $__________
The mail is full of these bills and checks: 150, $^{-}35.25$, 68.50, $^{-}14$.

Wednesday: beginning balance: $__________
Today, you receive 29.95, $^{-}19$, $^{-}17.80$.

Thursday: beginning balance: $__________
"Thursday's child is full of woe;" you get only bills: $^{-}29$, $^{-}57.18$, $^{-}56.60$.

Friday: beginning balance: $__________
"But Friday's child has far to go!"
You receive 80, 116.75, $^{-}13.80$.

The week's ending balance: $__________

Week 2

Monday: beginning balance: $ 200.00
The postman brings checks for 25.50 and 81.70, and a bill for $^{-}19.75$.

Tuesday: beginning balance: $__________
In the mail, you see only bills: $^{-}28.31$, $^{-}45.16$, $^{-}27$.

Wednesday: beginning balance: $__________
The postman brings better news today—checks for 50.75, 29.95, and only one $^{-}18.25$ bill.

Thursday: beginning balance: $__________
Another "Black" Thursday—you would end up in the "red"— $^{-}121.50$, $^{-}175.67$, $^{-}21.40$.

Friday: beginning balance: $__________
Luckily, Friday's mail brings only good news: 75, 99.50, 15.75, 88.57.

The week's ending balance: $__________

The Postman Always Rings Twice

Name ______________________________

Subtracting rational integers

Solve the problems below by subtracting decimal integers.

1. $0.3 - 0.145 =$
2. $^{-}0.7 - 0.365 =$
3. $^{-}1 - {}^{-}0.5461 =$
4. $5 - 1.3609 =$
5. $^{-}0.203 - {}^{-}14.1 =$
6. $0.3131 - 26 =$
7. $4.38 - {}^{-}8.756 =$
8. $^{-}1.2 - {}^{-}0.52 =$
9. $^{-}9.7 - {}^{-}52 =$
10. $^{-}8.7 - 973.647 =$
11. $0.03 - 0.00987 =$
12. $^{-}1 - {}^{-}0.983 =$
13. $^{-}57 - {}^{-}0.00101 =$
14. $0.019 - 0.2 =$
15. $^{-}5.7 - 61.823 =$
16. $603.485 - {}^{-}29.98 =$
17. $0.0231 - 3.3 =$
18. $^{-}2.9 - {}^{-}1.0407 =$
19. $14.71 - 65.2 =$
20. $^{-}18.486 - {}^{-}29 =$

Opposites Attract!

Multiplying rational integers

Name ______________________________

Solve each problem below by multiplying decimal integers.

1. $^{-}0.9(^{-}0.081)$ =
2. $0.14 \times {^{-}16}$ =
3. $(^{-}5.75)(0.067)$ =
4. $0.0036 \times {^{-}7.5}$ =
5. $(^{-}0.2)^3$ =
6. $(^{-}0.08)^2$ =
7. $^{-}0.7 \times {^{-}0.09} \times 8$ =
8. $(^{-}0.1)^5$
9. $(0.825)(^{-}3.49)$ =
10. $^{-}19.78 \times {^{-}0.586}$ =
11. $(^{-}0.013)^3$ =
12. $(^{-}2.5)^4$ =
13. $(0.029)(^{-}0.0057)$ =
14. $(^{-}0.0048)(0.875)$ =
15. $(^{-}678)(0.04)(^{-}0.15)$ =
16. $(^{-}0.01)^4$ =
17. $(^{-}0.5)^3$ =

Division of Labor

Dividing rational integers

Name ______________________________

Divide each problem until the number ends or repeats.

1. $\frac{^-3}{0.3} =$
2. $\frac{4.95}{^-15} =$
3. $\frac{^-36.18}{1.8} =$
4. $\frac{^-95}{^-0.005} =$
5. $\frac{0.00564}{^-4} =$
6. $\frac{^-97.8}{^-0.027} =$
7. $\frac{121}{^-1.1} =$
8. $\frac{^-8.34256}{^-0.736} =$
9. $\frac{^-14.3352}{5.43} =$
10. $\frac{0.00648}{^-7.42} =$
11. $\frac{0.00625}{^-25} =$
12. $\frac{^-39}{^-0.0013} =$
13. $\frac{^-40.96}{0.512} =$
14. $\frac{40}{^-0.003} =$
15. $\frac{^-0.003264}{16} =$
16. $\frac{2.5725}{0.343} =$
17. $\frac{^-100}{0.6} =$
18. $\frac{^-0.1}{^-9} =$

Monu-mental Math

Solving multi-type problems mentally

Name ______________________________

Solve each problem mentally, without pencil and paper.

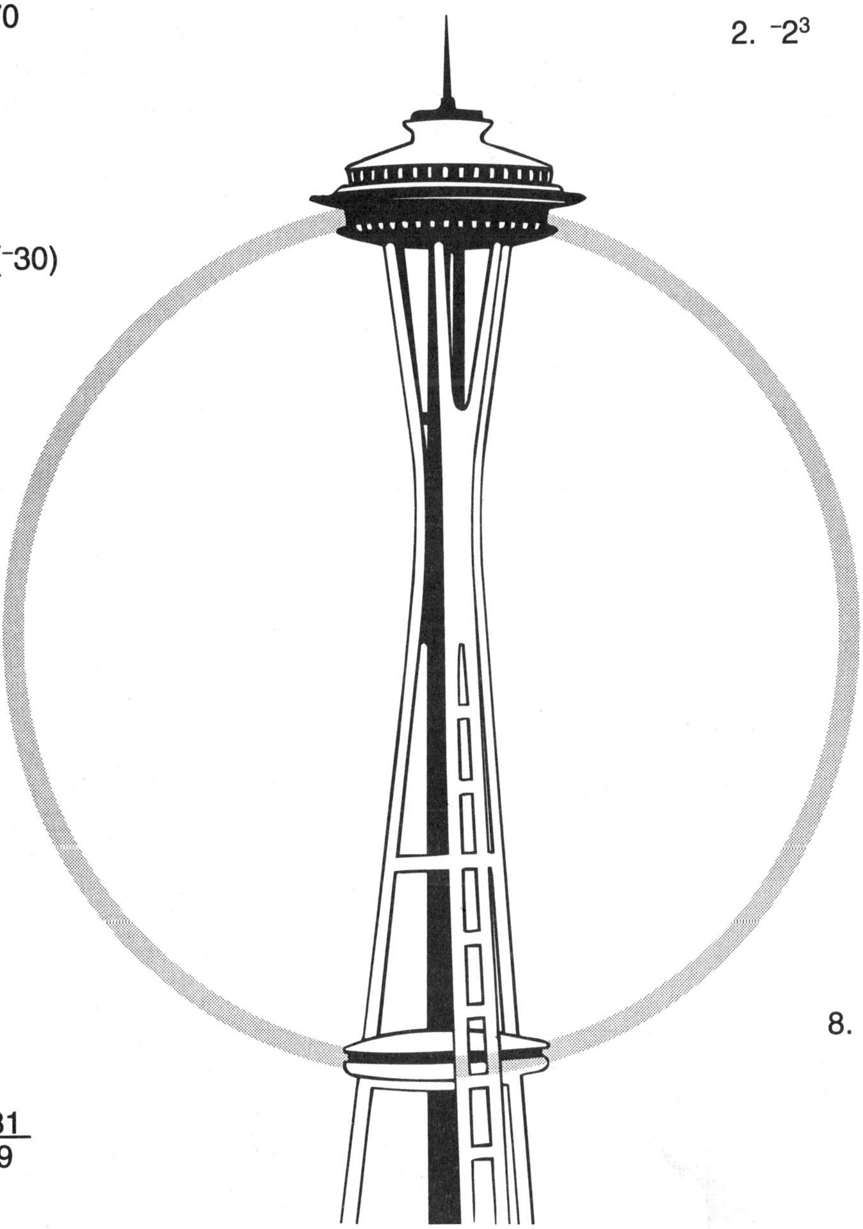

1. 50% of 70

2. $^{-}2^3$

3. $^{-}60(^{-}30)$

4. $5x = 45$

5. $^{-}5^2$

6. $18 - 21$

7. $\frac{1}{2} \bullet \frac{3}{8}$

8. 25% of 36

9. $\frac{^{-}81}{^{-}9}$

10. $^{-}3 + 5$

11. $x - 9 = 12$

12. $\frac{x}{9} = 8$

Equation Munchers

Multiple-step equations

Name ______________________________

Solve each equation for x. Then use the code to spell the name of a favorite food of many people. (Code—a = 0, b = ±1, c = ±2, d = ±3, etc.)

1. $\frac{30}{x} - 5 = {}^{-}3$
2. $15 - (x + 7) = 4$
3. $\frac{(x + 5)}{4} = 5$
4. $\frac{x}{3} + {}^{-}7 = {}^{-}12$
5. $2x + 7 = 3x + 3$
6. $45 - 3x = {}^{-}6$
7. $\frac{20 + x}{17} = 16 - x$
8. $2x + 4 = 36$
9. $5x + 10 = 7x - 6$
10. $\frac{x}{5} - 7 = {}^{-}4$
11. $\frac{30}{x + 2} = 3$
12. $100 - x = 3x$
13. $\frac{x}{5} + 5 = 0$
14. $9x + 7 = 7$

___	___	___	___	___	___	___	___	___		___	___	___	___	___
1	2	3	4	5	6	7	8	9		10	11	12	13	14

Olympic Triathlon

Fractional percents

Name ____

In the three "events" below, you must find the missing percent, decimal, or lowest-term fraction for each problem. Go for the gold!

	Percent	Decimal	Lowest-term fraction
1.	$\frac{1}{2}\%$	________	________
2.	________	________	$\frac{3}{8}$
3.	________	0.0025	________
4.	$8\frac{1}{4}\%$	________	________
5.	________	0.0625	________
6.	________	0.512	________
7.	________	________	$1\frac{4}{7}$
8.	$\frac{2}{3}\%$	________	________
9.	________	________	$2\frac{1}{6}$
10.	$3\frac{5}{8}\%$	________	________
11.	________	0.8333 . . .	________
12.	________	________	$\frac{1}{12}$
13.	$10\frac{3}{4}\%$	________	________
14.	________	0.666 . . .	________
15.	________	________	$\frac{1}{200}$

Hot Wok

Open-ended word problems with decimal, fractional percents

Name ______________________________

Appetizers		Beef		Chicken		Pork		Drinks	
4 wontons	3.95	Hot garlic beef	6.95	Almond chicken	4.95	Moo shu pork	6.75	Soft drink	0.75
5 fried shrimp	5.95	Moo shu beef	6.75	Lemon chicken	4.75	Sweet/sour pork	5.50	Tea	0.50
2 egg rolls	2.50	Beef/snow peas	4.75	Chicken/broccoli	5.25	Hot garlic pork	5.95	Coffee	1.00
6 rumaki	6.50	Beef/broccoli	5.25	Hot/spicy chicken	5.95	Pork/vegetables	4.75	Iced tea	1.50

Seafood		Noodles/Rice		Vegetables/Soups		Desserts	
Hot/spicy shrimp	5.95	Vegetable chow mein	2.95	Steamed platter	4.75	Ice cream	1.35
Hot garlic scallops	6.75	Crispy noodles	3.50	Broccoli/mushroom	2.50	Sherbet	1.35
Lobster/vegetables	8.50	Steamed rice	0.50	Snow peas/mushroom	2.50	Chocolate chip cookie	0.50
Sizzling seafood	8.50	Fried rice	1.50	Wonton soup	1.50	Almond cookie	0.35
Shrimp/snow peas	6.50	Pork fried rice	2.50	Sweet/sour soup	1.75	Fortune cookie	0.25

Solve the problems below on another sheet of paper.

1. You and your three friends are going to a Chinese restaurant. Use the menu items above to order a selection of items. Total your bill. Compute and add to your bill $8\frac{1}{4}\%$ tax and 15% tip.

   ____

2. Now, order a different selection of items for a dinner for a family of two adults and three children. Compute a $7\frac{1}{2}\%$ tax and add a 10% tip to get the final total bill.

   ____

3. Coupon Special! 25% off Hot Wok Meal! After ordering a complete meal for three people consisting of appetizer, three entrees, drinks, and dessert, use the coupon to take 25% off the sub-total bill. Then calculate an $8\frac{3}{4}\%$ tax and a 15% tip to get the final amount.

   ____

4. Hot Wok also offers a catering service. Order a complete banquet for your family reunion—estimate about 25–30 people! Total the bill. Calculate an 8% tax and a 15% tip.

   ____

5. The sky is the limit! For a very special 20th wedding anniversary for your parents, you and your brother are treating them to the very best selections. Calculate the total bill. Don't forget an $8\frac{3}{4}\%$ tax and a 15% tip.

   ____

It's My Bag

Word problems with sale items

Name ____

Today is SUPER SALE DAY at the mall, and you are going on a shopping spree.

1. Bathing suit—30% off, original price $54. You pay how much? __________
2. Tennis shoes—advertised at 25% off, you paid $30. They were originally how much? __________
3. Jeans were discounted from $35 to $25. What was the percent of discount? __________
4. Big T-Shirt Sale! Three for $25, originally $10 each. What was your percent of savings? __________
5. Discontinued brand shirt sale—70% off original price of $35.50, now priced to sell at how much? __________
6. Leather jackets reduced 20%. Now selling for $64, their former price was how much? __________
7. Coats that had been priced at $65 were reduced to sell at $52. What was the percent of discount? __________
8. You couldn't decide! One top was marked down 25% from $30; the other was 20% off $28. Which one cost less? __________
9. What was the total amount you spent today? __________
10. What was the total amount and the percent you saved? __________

____

Who Owns What?

Logic

Name ______________________________

Five friends live on the same street in different colored houses. They each have a different pet and mode of transportation. Use the clues to fill in the puzzle.

1. Pablo lives in the red house.
2. Maria often plays after school with her dog.
3. The boy who runs owns the cat.
4. The bicycle rider lives in the yellow house with the goldfish.
5. David's hamster loves the blue house where it lives.
6. The walker loves her parrot.
7. The boy who owns the hamster skates.
8. The girl who owns the parrot lives in the green house.
9. Lynn walks to school.

	House					Pet					Mode of Transportation				
	purple	yellow	blue	red	green	goldfish	hamster	cat	dog	parrot	bikes	skates	runs	skate-boards	walks
Pablo															
Maria															
David															
Lynn															
Beth															

Pablo = ______________________________

Maria = ______________________________

David = ______________________________

Lynn = ______________________________

Beth = ______________________________

Birthday "Day"-ta

Organizing and displaying data

Name ______________________________

In a survey of birthday months in their eighth-grade homeroom, the students found the months listed below as their "raw data". First, organize the data into a T-chart. Then display the information in a bar graph. Identify the mode and any gaps in the data. Write a statement about your findings.

September, September, July, April, June, August, December, October, July, January, November, March, November, September, April, June, January, August, July, November, March, January, November, June, April, July, September, August, March, June, November, October

T-chart

Mode______________________________

Gaps______________________________

Findings ______________________________

____

____

Bar Graph

Mathemania Crossword

Name ______________________________

Complete the crossword puzzle below.

Across

1. a liquid measure
4. greater than 90°
5. 90°
6. horizontal circles
10. the point
11. π
12. length
13. less than 90°
14. positive and negative numbers
16. side of an angle
18. ray intersection
20. half a diameter
21. never intersecting

Down

1. _____ theorem
2. intersecting at 90° angle
3. distance around a circle
7. 2 times radius
8. more than one radius
9. dollars and _____
15. metric measurement of length
17. x or y line
19. =

Pizza Pi

Area and circumference of a circle

Name ____

Find the area and circumference of each pizza. Use $\pi \approx 3.14$ or $3\frac{1}{7}$. (Round to nearest $\frac{1}{100}$.)

$$A = \pi r^2 \qquad C = \pi d$$

1. Small mushroom and sausage for one person with a diameter of 7"

2. Medium pepperoni for two with a radius of 5"

3. Individual combo with a radius of 4"

4. Large combo special with pepperoni, sausage, onion, bell pepper, mushrooms, and extra cheese—15.5" diameter

5. The "Whopper With the Works" with a $10\frac{1}{2}$" radius

6. Extra large "Veggie Special" with a 16" diameter

7. If $\frac{1}{4}$ of a 14" diameter pizza had been eaten, what would the remaining perimeter be? Is it more or less than the original circumference?

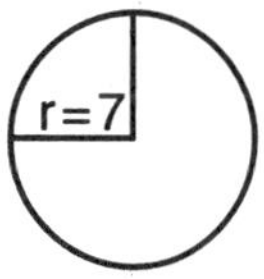

8. You and your three friends are equally splitting one medium pepperoni pizza having a 10" diameter. What is the measure of the arc of each slice? What is the area of each slice?

9. Which has the greater area—2 medium sausage and mushroom pizzas, each with a 6" radius, or one large sausage and mushroom pizza with a circumference of 56.52"?

10. Two competing pizzerias offer these two different specials: Pizza Pete's has a large 8" radius "Peterworks" for $8.99; Plumbers Pizza Parlor advertises an "All But the Kitchen Sink" special, with a circumference of 50.24" for the same $8.99 price. Which is the better deal?

It's a Square World, After All

Simplifying radicals and estimating square roots

Name ____

Simplify each radical by prime factorization or by multiplying radicals.

1. $\sqrt{20}$ =
2. $\sqrt{18}$ =
3. $\sqrt{12}$ =
4. $\sqrt{40}$ =
5. $\sqrt{75}$ =
6. $\sqrt{300}$ =
7. $\sqrt{140}$ =
8. $\sqrt{625}$ =
9. $\sqrt{500}$ =
10. $2\sqrt{5} \cdot 2\sqrt{5}$ =
11. $3\sqrt{3} \cdot \sqrt{3}$ =
12. $\sqrt{512}$ =

Solve each radical by the averaging method of estimating square roots. Round the answers to the nearest hundredth.

13. $\sqrt{12}$ =
14. $\sqrt{20}$ =
15. $\sqrt{115}$ =
16. $\sqrt{75}$ =
17. $\sqrt{29.5}$ =
18. $\sqrt{70.6}$ =
19. $\sqrt{0.02}$ =
20. $\sqrt{0.18}$ =

Right On!

Area and perimeter of right triangles

Name ______________________________

Solve each right triangle problem for area and perimeter. Use $a^2 + b^2 = c^2$, the averaging method to estimate square roots (round to the nearest tenth), $A = \frac{1}{2}bh$, and $P = 2a + b$ or $P = a + b + c$.

1.
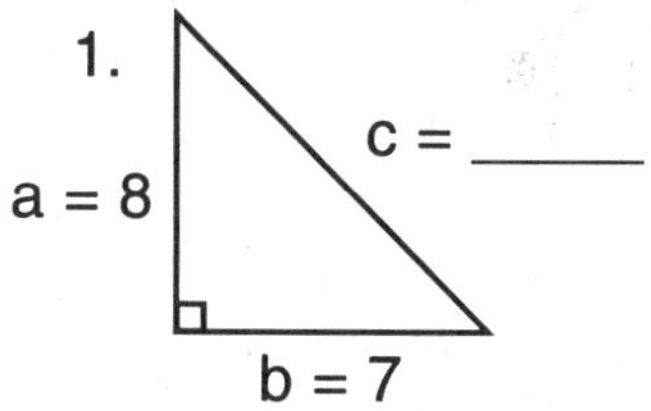

2.
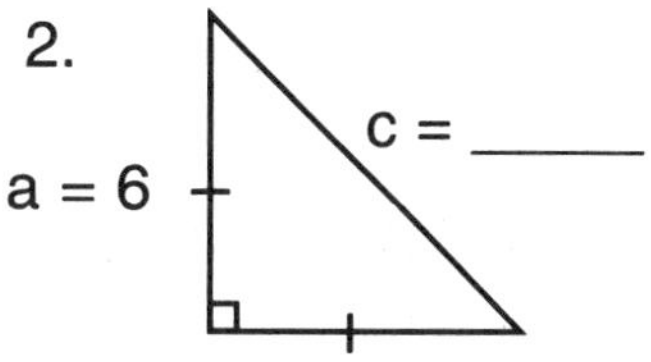

3.
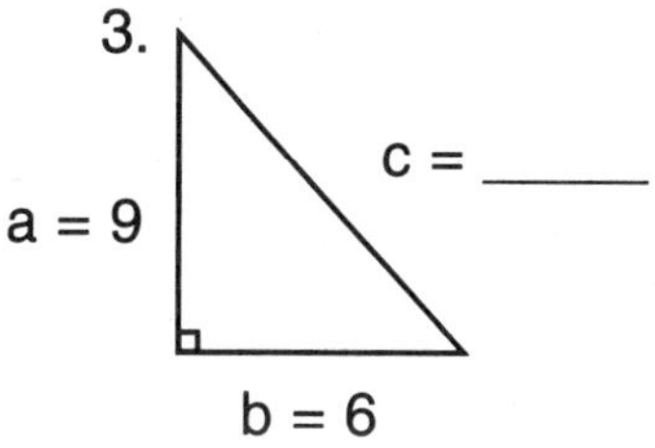

4.
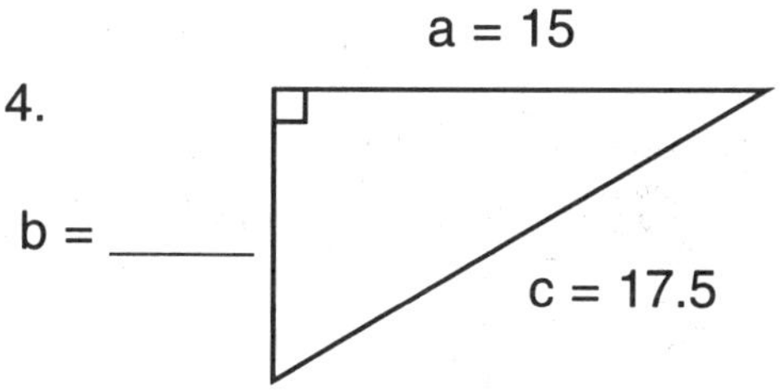

5.
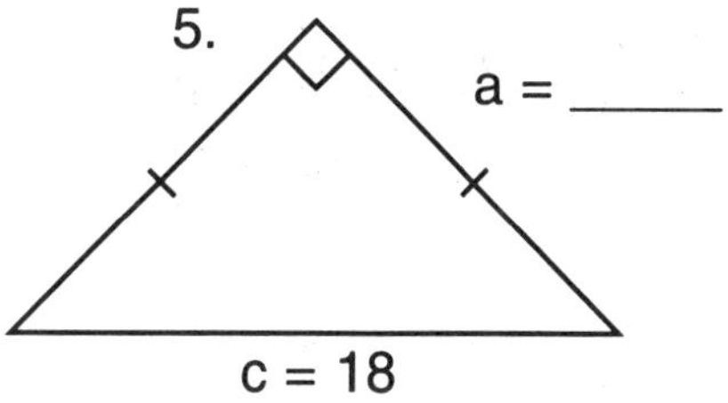

6.
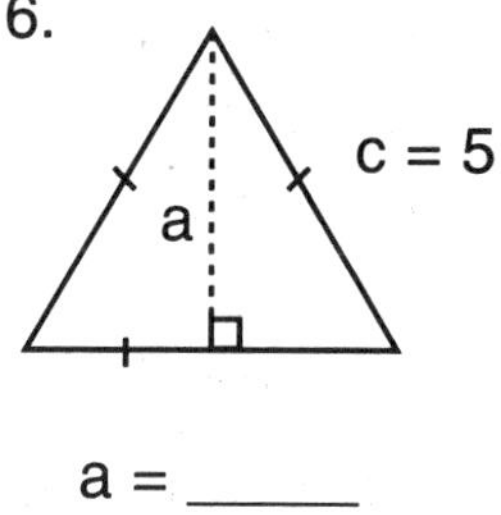

7.
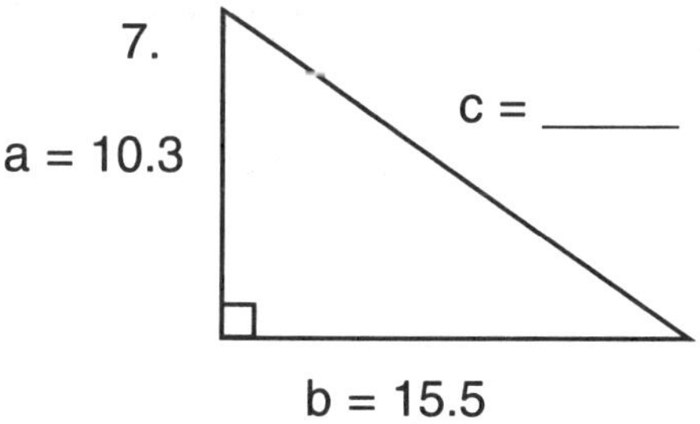

8.
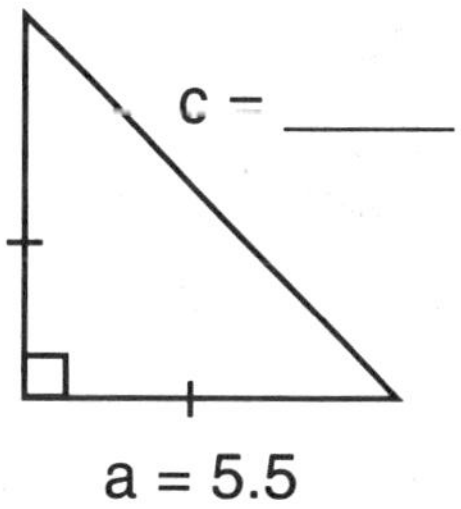

9.
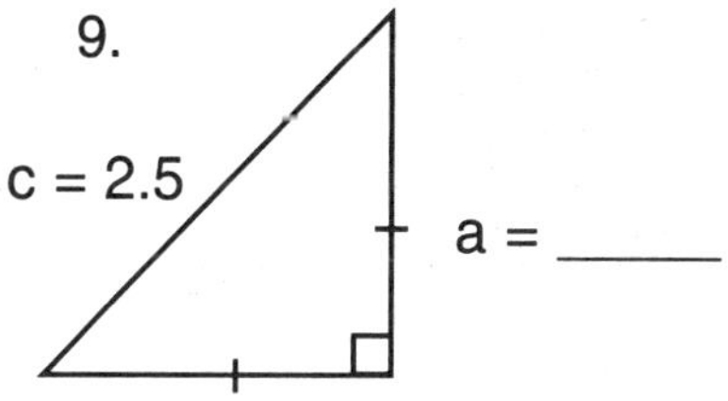

10.
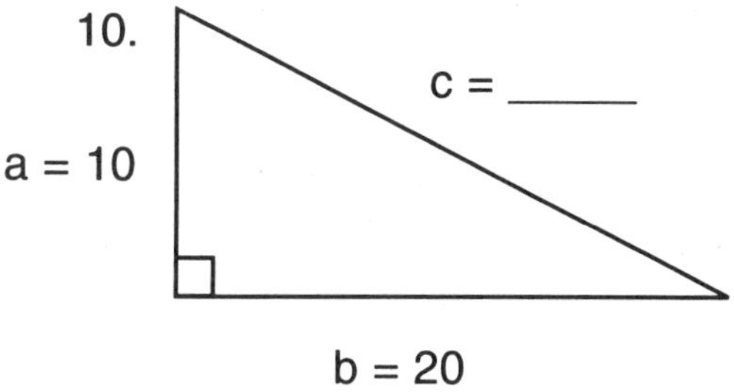

11.
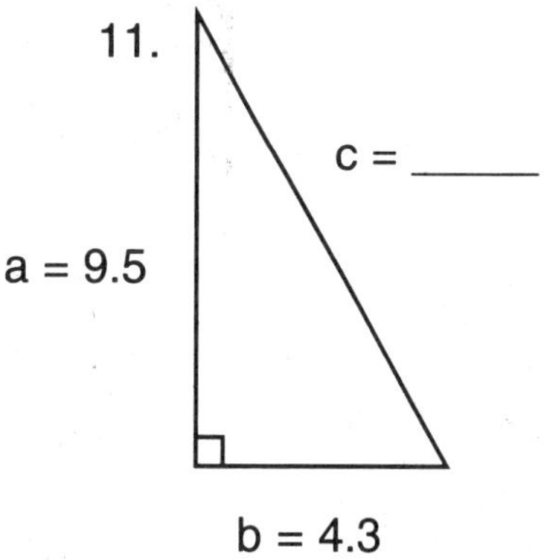

Trap a Zoid

Area of trapezoids

Name ________________________________

You have a mission. Solve each trapezoid's area to find the combination that unlocks the door to the gorilla's cage so that it is free. Use $A = \frac{1}{2}h(b_1 + b_2)$.

ABCD = __________

AEFG = __________

GHIJ = __________

IKLM = __________

KLNO = __________

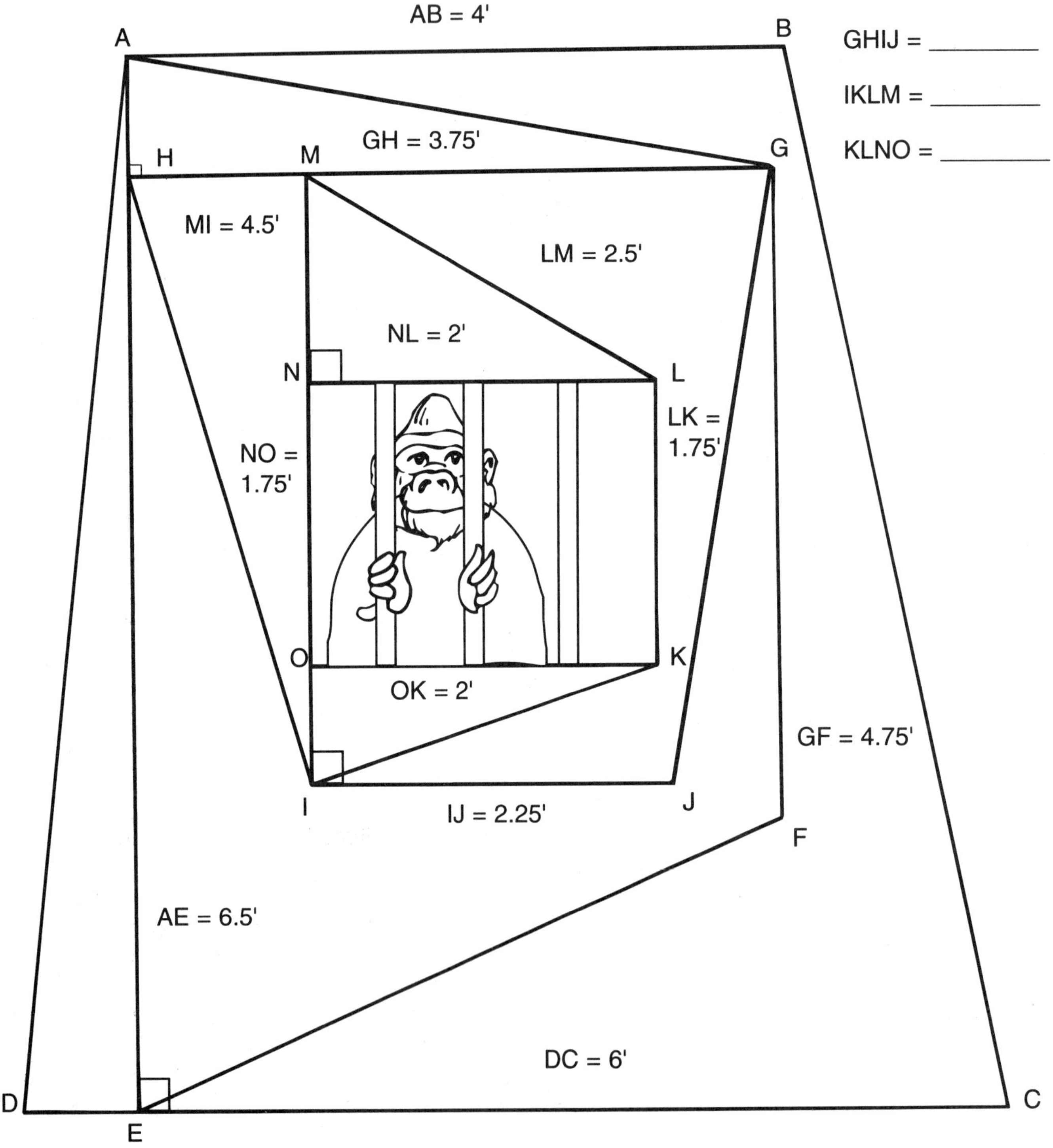

Luxor Revisited

Volume and surface area of square pyramids and tetrahedrons

Name ____

Solve each problem below.

1. Square pyramid whose base has each side (s) = 4', altitude height = 2.8', slant height = 3.5'. Use volume (V) = ⅓Bh, with h = alt. height; surface area (SA) = $s^2 + 4(\frac{1}{2}bh)$, with h = slant height.

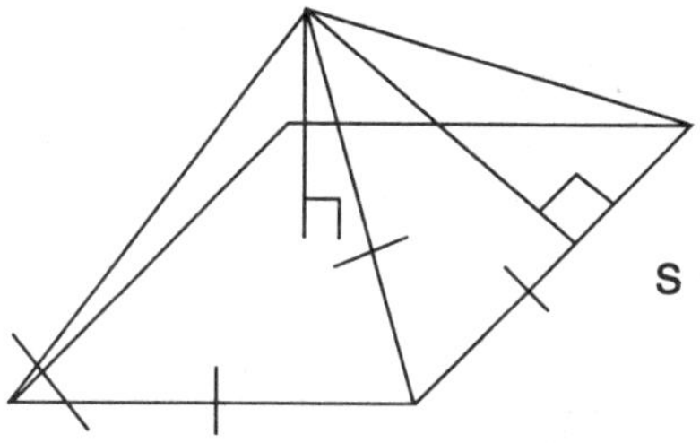

2. Tetrahedron with all sides (s) = 10 cm. SA = $4(\frac{1}{2}bh)$ with slant height = 8.7 cm.

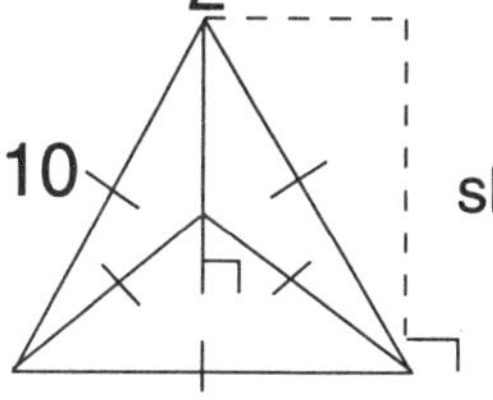

*3. Given a square pyramid with s = 12". Use $a^2 + b^2 = c^2$ to calculate slant height and altitude height. Then find the volume and surface area.

*4. This tetrahedron has an alt. height = 28 mm and a slant height = 35 mm. Use the Pythagorean theorcm to find the length of the side (s). Then calculate volume and surface area. (Hint: Use $c^2 = s^2$, $a^2 = (.5s)^2$.)

5. Given a rectangular pyramid with a base of ℓ = 10 m and w = 8 m, find V and SA if the altitude height = 3 m and slant height = 5 m. Use SA = B + $2(\frac{1}{2}wh) + 2(\frac{1}{2}\ell h)$, where h = slant height.

*6. Given a tetrahedron with all faces (s) = 8', find slant height, altitude height, V, and SA.

7. Given a square pyramid with s = 24", alt. height = 5", and slant height = 13", find V and SA.

*8. Given a square pyramid with an alt. height = 10 cm, and slant height = 15 cm, find the length of the side of the square base. Then find V and SA.

Pi Makes the World Go 'Round

Volume and surface area of cylinders, cones, and spheres

Name ___________________________________

Find volume and/or surface area for each of the figures below. Use $\pi \approx 3.14$ or $3\frac{1}{7}$.

1. Find volume and surface area of a Ferris wheel if its radius equals 10' and the width of the seats is 4'. Use the formulas $V = \pi r^2 h$ and $SA = 2\pi r^2 + Ch$.

2. Find the volume of Nancy's snow cone before it melts! Use d = 2" and h = 6". For the top ½ sphere, $V = \frac{1}{2}(\frac{4}{3}\pi r^3)$; for the bottom cone, $V = \frac{1}{3}\pi r^2 h$.

3. Jim's basketball has a diameter of 10". What is its volume and surface area? Use the formulas $V = \frac{4}{3}\pi r^3$ and $SA = 4\pi r^2$.

*4. The circumference of Earth at its equator is approximately 25,000 miles. Find its approximate volume and surface area. Use $d = \frac{25{,}000}{\pi}$. Round to the nearest hundredth.

5. Marilyn and Annette said their pizza was more than the volume of one of their 3" x 15" x 10" desktops. If they had ordered the 3" thick "Colossal Deluxe" with a 15" diameter, were they right or wrong?

6. One of the "coneheads" wanted to measure the size of his head. He found that his circumference was 21", and the height was 10". Help him find the volume.

7. A portable cylindrical swimming pool in Manny and Moe's backyard is 10' across and 6' deep. If they fill it to the rim, what is the volume of water needed?

8. Marcy and Darcy were training for the Olympics in their mini cylindrical pool. If its volume was 1,000 cu. ft., and it was 10' deep, how many laps would they each need to do to swim ½ mile per day? (Use 1 mi = 5,280 ft.) Consider that one lap consists of swimming across and back the widest part of the pool. Round r^2 to the nearest whole number.

Under Construction

Non-metric constructions of geometric figures

Name ______________________________

Using a compass and a straight edge, construct the figures below.

1. Copy ∠ABC. Name the new angle ∠DEF.

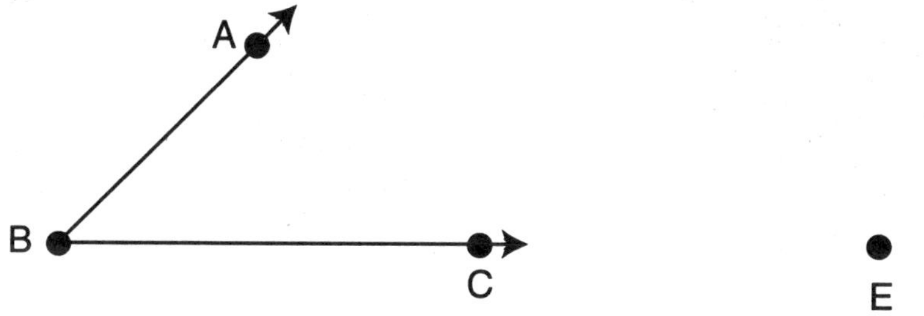

2. Bisect ∠XYZ.

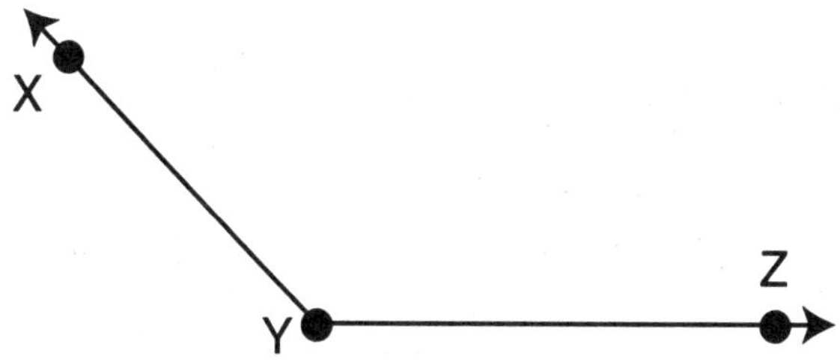

3. Construct the perpendicular bisector of line segment AB.

4. Construct line MN parallel to line CD using point-not-on-a-line method.

5. Construct right isosceles triangle ABC using point-on-a-line method with each leg congruent to 1".

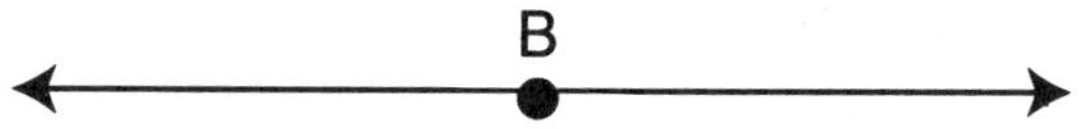

Fantastic Functions

First-degree function T-charts

Name ______________________________

Make a T-chart for each function using x = $^{+}5$ to $^{-}5$ values. Find the repeating differences between the y values for each function.

1. $y = x + {}^{-}3$
2. $y = x - {}^{-}9$
3. $y = 2x + 1$
4. $y = 2x - 5$
5. $y = \frac{x}{2}$
6. $y = {}^{-}x + 2$
7. $y = 3x + 6$
8. $y = {}^{-}2x - 1$
9. $y = 3x + 3$
10. $y = \frac{{}^{-}x}{2} + 1$
11. $y = 5x + 5$
12. $y = 4x - 4$

Graph It!

Graphing first-degree functions

Name ______________________________

Make a T-chart for each function for x = ⁺5 to ⁻5 values. Graph as far as y = ⁺20 or ⁻20. You will need one-fourth inch graph paper. You may put up to three lines per graph. Name each line.

1. $y = x + {}^{-}2$
2. $y = x - {}^{-}7$
3. $y = 2x - 1$
4. $y = 2x + 4$
5. $y = \frac{x}{2}$
6. $y = {}^{-}x + 3$
7. $y = 3x + 5$
8. $y = {}^{-}2x - 2$
9. $y = 3x + {}^{-}2$
10. $y = \frac{{}^{-}x}{2} + 1$
11. $y = 5x - 5$
12. $y = 4x + 4$

How Many Squares?

Visual spatialization

Name ______________________________

1. How many squares are there on a standard checkerboard? While many of the squares are obvious, many are not. Remember to look for squares of various sizes. Also, be ready to explain your count of the squares.

   Number of Squares ________

2. How many triangles can you find in the figure below? Count carefully!

   Number of Triangles ________

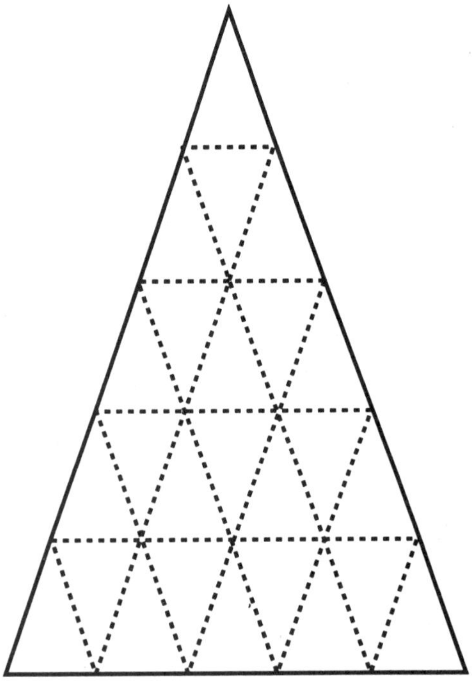

3. How many squares can you find in the figure to the right? Count carefully!

   Number of Squares ________

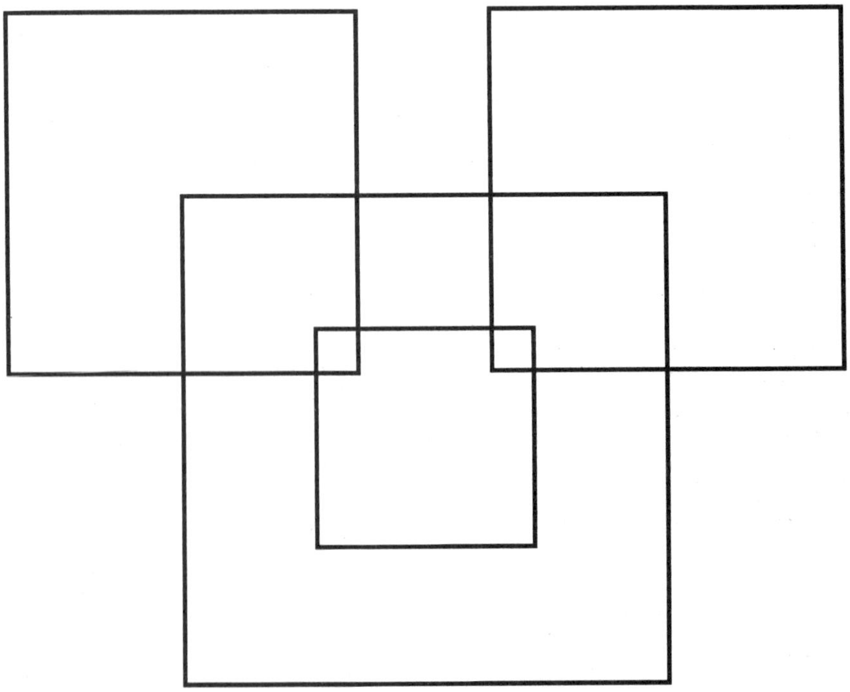

Answers

Page 2

A. 1st $^-16$; 2nd 3; 3rd 4; Home $^-6$
B. 1st $^-2$; 2nd $^-3$; 3rd 0; Home $^+5$
C. 1st $^-1$; 2nd $^-1$; 3rd $^-2$ 1/2; Home $^-6$

Page 3

I. 1. 2 5/7; 2. 2 2/3; 3. 1 3/4; 4. 1 3/5; 5. 0; 6. $^-$4/7; 7. $^-$5/8; 8. $^-$2/3; 9. $^-1$ 1/3; 10. $^-1$ 3/8
II. 1. 1 2/3; 2. 1 3/5; 3. 5/9; 4. 1/2; 5. 0; 6. $^-$5/16; 7. $^-$3/8; 8. $^-$4/9; 9. $^-1$ 1/3; 10. $^-1$ 2/5
III. 1. 8/9; 2. 7/8; 3. 13/16; 4. 4/5; 5. 0; 6. $^-1$ 9/20; 7. $^-1$ 1/2; 8. $^-2$ 3/16; 9. $^-2$ 2/9; 10. $^-2$ 1/4

Page 4

A. 1. $^-9$ 7/20; 2. $^-18$ 9/40; 3. $^-15$ 29/40; 4. $^-8$ 97/120; 5. $^-9$ 41/60; 6. 5 14/15
B. 1. $^-$1/6; 2. $^-8$ 11/12; 3. $^-12$ 7/12; 4. $^-6$ 23/24; 5. $^-5$ 11/24; 6. 2 19/24

Page 5

1. $^-2$ 1/2 2. $^-3$ 1/4 3. $^-16$ 1/6
4. $^-15$ 27/28 5. 7 13/36 6. 36 7/12
7. $^-10$ 3/8 8. $^-6$ 3/7 9. $^-27$ 17/18
10. $^-6$ 23/24 11. $^-1$ 13/60 12. $^-18$ 1/20
13. $^-3$ 11/24 14. $^-29$ 2/3

Page 6

1. 9 11/10, 5 2/5, 2 7/12, 3 19/20, 2 1/2, 2 3/8
2. 1 7/8, 2 5/12, 5 5/8, 2 5/6, 9 7/8, 11 2/9
3. 7 7/9, 13 5/6, 4 3/5, 8 7/8, 1 23/36, 11 57/88
4. 47/48, 3 14/15, 13 23/30, 5 6/7, 16 31/42, 3 8/9

Page 7

1. $^-$4/7 2. $^-$1/3 3. 9/16
4. $^-$16/135 5. $^-$8/125 6. 1/256
7. $^-5$ 8. 20 9. 51
10. $^-16$ 11. $^-21$ 1/3 12. 2 7/9
13. 16 2/3 14. $^-6$ 15. $^-20$ 51/64
16. 39 1/16

Page 8

A BUNCH OF BANANAS

Page 9

Week 1: Tues. $^-75.67$; Wed. 93.58; Thurs. 86.73; Fri. $^-56.05$; Ending: $126.90
Week 2: Tues. 287.45; Wed. 186.98; Thurs. 249.43; Fri. $^-69.14$; Ending: $209.68

Page 10

1. 0.155 2. $^-1.065$ 3. $^-0.4539$
4. 3.631 5. 13.897 6. $^-25.6869$
7. 13.136 8. $^-0.68$ 9. 42.3
10. $^-982.347$ 11. 0.02013
12. $^-0.017$ 13. $^-56.99899$
14. $^-0.181$ 15. $^-67.523$
16. 633.465 17. $^-3.2769$
18. $^-1.8593$ 19. $^-50.49$
20. 10.514

Page 11

1. 0.0729 2. $^-2.24$
3. $^-0.38525$ 4. $^-0.027$
5. $^-0.008$ 6. 0.0064
7. 0.504 8. $^-0.00001$
9. $^-2.87925$ 10. 11.59108
11. $^-0.000002197$
12. 39.0625 13. $^-0.0001653$
14. $^-0.0042$ 15. 4.068
16. 0.00000001 17. $^-0.125$

Page 12

1. $^-10$ 2. $^-0.33$
3. $^-20.1$ 4. 19,000
5. $^-0.00141$ 6. 3622.2 . . .
7. $^-110$ 8. 11.335
9. $^-2.64$ 10. $^-0.0008733$ . . .
11. $^-0.00025$ 12. 30,000
13. $^-80$ 14. $^-13,333.3$ . . .
15. $^-0.000204$ 16. 7.5
17. $^-166.6$ . . . 18. 0.011 . . .

Page 13

1. 35 2. $^-8$ 3. 1800
4. x = 9 5. 25 6. $^-3$
7. 3/16 8. 9 9. 9
10. 2 11. x = 21 12. 72

Page 14

1. 15 2. 4 3. 15 4. $^-15$
5. 4 6. 17 7. 14 8. 13
9. 8 10. 15 11. 8 12. 25
13. $^-25$ 14. 0
pepperoni pizza

Page 15

1. 0.005, 1/200
2. 37 1/2%, 0.375
3. 1/4%, 1/400
4. 0.0825, 33/400
5. 6 1/4%, 1/16
6. 51 1/5%, 64/125
7. 157 1/7, 1.571428 . . .
8. 0.006666 . . ., 1/150
9. 216 2/3%, 2.1666 . . .
10. 0.03625, 29/800
11. 83 1/3%, 5/6
12. 8 1/3%, 0.0833 . . .
13. 0.1075, 43/400
14. 66 2/3%, 2/3
15. 0.5%, 0.005

Page 16

Answers will vary.

Page 17

1. $37.80 2. $40 3. 28.6%
4. 16.7% 5. $10.65 6. $80
7. 20% 8. 20% of $28 = $22.40
9. $266.85 10. $100.65, 27.4%

Page 18

Pablo—red house, cat, runs
Maria—purple house, dog, skateboards
David—blue house, hamster, skates
Lynn—green house, parrot, walks
Beth—yellow house, goldfish, bikes

Page 19

Month	#
Jan.	III
Feb.	
Mar.	III
April	III
May	
June	IIII
July	IIII
Aug.	III
Sep.	IIII
Oct.	II
Nov.	~~IIII~~
Dec.	I

Mode = November
Gaps = Feb.; May
Findings will vary.

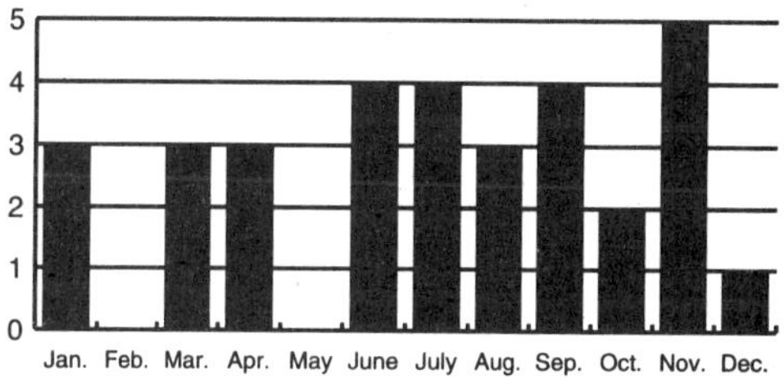

Page 20

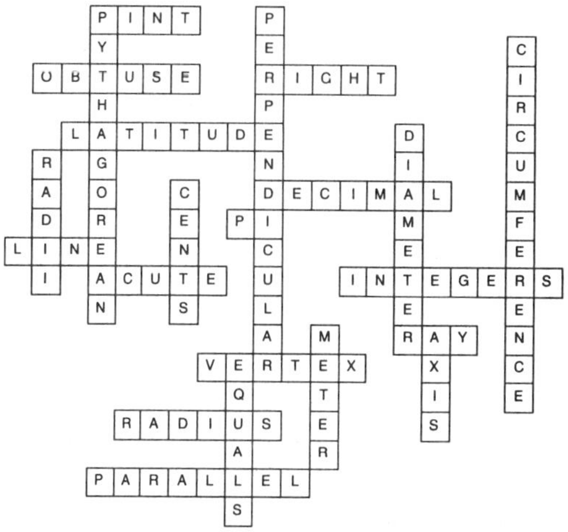

Answers

Page 21

1. A = 38.47", C = 21.96"
2. A = 78.5", C = 31.4"
3. C = 25.12", A = 50.24"
4. A = 188.6", C = 48.67"
5. A = 346.19", C = 65.94"
6. C = 50.24", A = 200.96"
7. P = 32.97", less
8. arc = 7.85", A = 19.63"
9. A = 226.08" for 2 medium pizzas; A = 254.34" for one large pizza—greater area
10. They are the same area, A = 200.96".

Page 22

1. $2\sqrt{5}$
2. $3\sqrt{2}$
3. $2\sqrt{3}$
4. $2\sqrt{2 \cdot 5}$
5. $5\sqrt{3}$
6. $10\sqrt{3}$
7. $2\sqrt{7 \cdot 5}$
8. 25
9. $10\sqrt{5}$
10. 10
11. 9
12. $16\sqrt{2}$
13. 3.46
14. 4.47
15. 10.72
16. 8.66
17. 5.43
18. 8.4
19. 0.14
20. 0.42

Page 23

1. c = 10.6, A = 28, P = 25.6
2. c = 8.49, A = 18, P = 20.49
3. c = 10.8, P = 25.8, A = 27
4. b = 9, A = 67.5, P = 41.5
5. a = 12.7, A = 80.6, P = 43.4
6. a = 4.3, P = 15, A = 10.75
7. c = 18.6, A = 79.8, P = 44.4
8. c = 7.8, P = 18.8, A = 15.1
9. a = 1.8, P = 6.1, A = 1.6
10. c = 22.4, P = 52.4, A = 100
11. c = 10.4, P = 24.2, A = 20.4

Page 24

ABCD = 32.5', AEFG = 21.1', GHIJ = 13.5', IKLM = 6.25', KLNO = 3.5'

Page 25

1. V = 14.9 cu. ft.; SA = 44 sq. ft.
2. SA = 174 cm^2
3. slant h = 10.4; alt. h = 8.5; V = 403.9 cu. in.; SA = 393.6 sq. in.
4. s = 40.4; V = 6532.7 mm^3; SA = 2828 mm^2
5. V = 79.2 m^3; SA = 170 m^2
6. alt. h = 3'; slant h = 5'; V = 20 cu. ft.; SA = 80 sq. ft.
7. V = 950.4 cu. in.; SA = 1200 sq. in.
8. s = 10 cm; V = 330 cm^3; SA = 175 cm^2

Page 26

1. V = 1256 cu. ft., SA = 879.2 sq. ft.
2. V = 8.37 cu. in.
3. V = 523.33 cu. in., SA = 314 sq. in.
4. V = 264,124,639,099.8 cu. mi., SA = 199,044,414.01 sq. mi.
5. Right; pizza = 529.8 cu. in., desktop = 450 cu. in.
6. V = 116.77 cu. in.
7. V = 471 cu. ft.
8. 110 laps each

Page 27

Check students' constructions.

Page 28

1.

x	y	d_1
5	2	1
4	1	1
3	0	1
2	-1	↓
1	-2	
0	-3	
-1	-4	
-2	-5	
-3	-6	
-4	-7	
-5	-8	

2.

x	y	d_1
5	14	1
4	13	1
3	12	1
2	11	↓
1	10	
0	9	
-1	8	
-2	7	
-3	6	
-4	5	
-5	4	

3.

x	y	d_1
5	11	2
4	9	2
3	7	2
2	5	↓
1	3	
0	1	
-1	-1	
-2	-3	
-3	-5	
-4	-7	
-5	-9	

4.

x	y	d_1
5	5	2
4	3	2
3	1	2
2	-1	↓
1	-3	
0	-5	
-1	-7	
-2	-9	
-3	-11	
-4	-13	
-5	-15	

5.

x	y	d_1
5	2.5	0.5
4	2	0.5
3	1.5	0.5
2	1	↓
1	0.5	
0	0	
-1	-0.5	
-2	-1	
-3	-1.5	
-4	-2	
-5	-2.5	

6.

x	y	d_1
5	-3	1
4	-2	1
3	-1	1
2	0	↓
1	1	
0	2	
-1	3	
-2	4	
-3	5	
-4	6	
-5	7	

7.

x	y	d_1
5	21	3
4	18	3
3	15	3
2	12	↓
1	9	
0	6	
-1	3	
-2	0	
-3	-3	
-4	-6	
-5	-9	

8.

x	y	d_1
5	-11	2
4	-9	2
3	-7	2
2	-5	↓
1	-3	
0	-1	
-1	1	
-2	3	
-3	5	
-4	7	
-5	9	

9.

x	y	d_1
5	18	3
4	15	3
3	12	3
2	9	↓
1	6	
0	3	
-1	0	
-2	-3	
-3	-6	
-4	-9	
-5	-12	

10.

x	y	d_1
5	-1.5	0.5
4	-1	0.5
3	-0.5	0.5
2	0	↓
1	0.5	
0	1	
-1	1.5	
-2	2	
-3	2.5	
-4	3	
-5	3.5	

11.

x	y	d_1
5	29	5
4	24	5
3	19	5
2	14	↓
1	9	
0	4	
-1	-1	
-2	-6	
-3	-11	
-4	-16	
-5	-21	

12.

x	y	d_1
5	15	4
4	11	4
3	7	4
2	3	↓
1	-1	
0	-5	
-1	-9	
-2	-13	
-3	-17	
-4	-21	
-5	-25	

Page 29

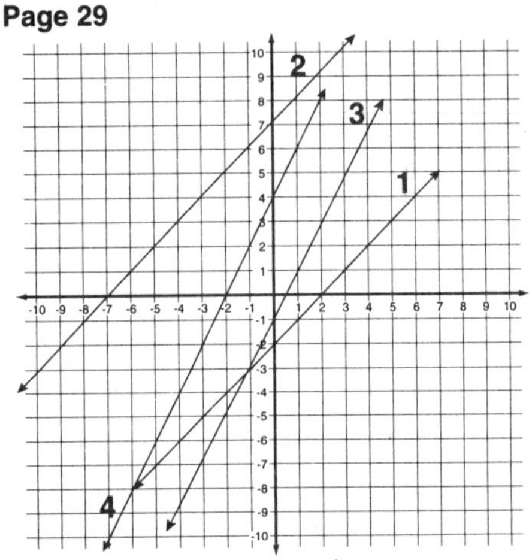

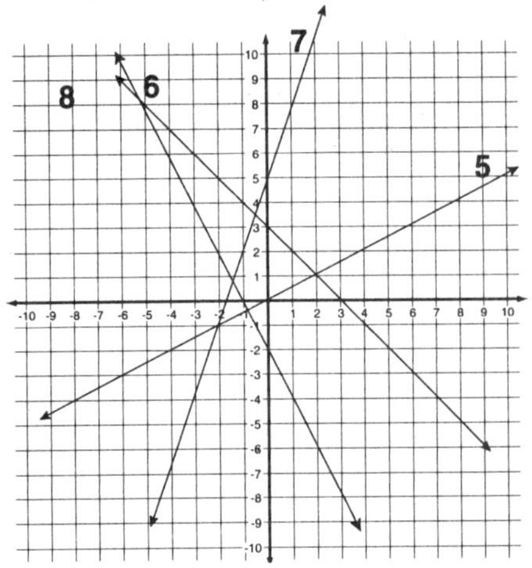

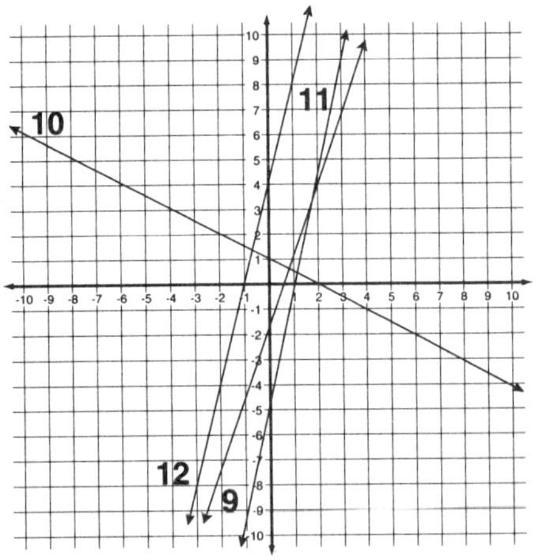

Page 30

1. There are 204 squares.
   64: 1 x 1; 49: 2 x 2; 36: 3 x 3; 25: 4 x 4; 16: 5 x 5; 9: 6 x 6; 4: 7 x 7; 1: 8 x 8
2. There are 47 triangles.
   25: 1 x 1; 13: 2 x 2; 6: 3 x 3; 2: 4 x 4; 1: 5 x 5
3. 9